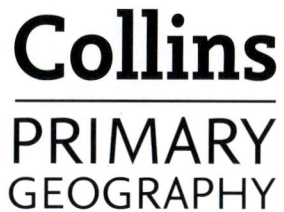

Our planet
Pupil Book 2

Earth in space	Earth, sun and moon	4
	The planets	6
	Day and night	8
	Land and water	10
Planet Earth	A living planet	12
	The shape of the land	14
	Volcanoes	16
	World wonders	18
Weather and seasons	Experiencing the weather	20
	Different types of weather	22
	Extreme weather	24
	The seasons	26
	Going round the sun	28
Local areas	Shelter	30
	Houses around the world	32
	Living in a village	34
	Exploring local streets	36
	Under your feet	38
Maps and plans	Maps and stories	40
	Treasure island	42
	Different plans	44
	The view from above	46
The United Kingdom	Countries and capitals in the United Kingdom	48
	Mountains, rivers and seas in the United Kingdom	50
Different environments	Living in the Arctic	52
	Living in the rainforest	54
	Living in the desert	56
	Animals around the world	58
World maps	World continents and oceans	60
	World countries	62

Stephen Scoffham | Colin Bridge

Our planet

Geography is about the world around us.

It's about volcanoes, mountains, rivers and deserts.

It's about people and the towns, villages and cities where they live.

Stories, pictures and maps show an amazing world.

Become a geographer.

Find out why things happen and how things change.

In this book you will meet lots of different characters. They will help you learn more about geography.

The Darin family: Mum, Baby Oscar, Ava and Theo;

Max and Azim;

Mika and Priya;

Toya and Efe;

Ewan and Isla;

Pia and Arjun;

Mrs Jones and her class;

Nimtok and Tami.

Earth, sun and moon

The sun and moon move through the sky above the Earth. The sun brings us warmth. The moon causes Earth's tides.

Read and talk about the story.

Ava's trick

Baby Oscar just couldn't sleep.
"Lie still and you will soon be asleep," said Mum.

In a little while, Baby Oscar came downstairs again.
"It's no use," he said. "I just can't sleep."

"Close your eyes and try again," said Mum.

"But I can't close my eyes," said Baby Oscar. "Ava told me that if I close my eyes, the big white ball in the sky will fall on me!"

"Ava is tricking you," laughed Mum. "The big white ball is the moon. It has been in the sky since time began."

What does Baby Oscar learn?

Read the words

Earth   sky
sun   tide
moon

Use the words to draw a picture or make a model. Write about it.

Amazing fact! There is red hot rock beneath the Earth's surface.

Talking
What are the differences between the Earth, sun and moon?

For you to do
Look at the moon as it gets dark. Is it always the same shape?

The planets

Planets are big balls of gas and rock spinning in space. The Earth is one of eight planets which go round the sun.

Read and talk about the story.

A journey in space

The Darin family set off on a journey in their space ship. There was Theo, Ava and Baby Oscar.

Baby Oscar was learning to count. As they left Earth, Ava said, "How many planets can you see between here and the sun?"

Oscar looked hard. "One, two," he said.

"Now add the Earth," said Theo.

"One, two, four," said Oscar.

"No, no!" laughed Ava. "One, two, three."

Soon they passed Mars and Jupiter.

"One, two, three, seven, nine," said Oscar.

"It's one, two, three, four, five," chuckled Theo.

Later Saturn, Uranus and Neptune went by.

"One, two, three, four, five, seven, six, eight," said Oscar.

"Try again," said Ava, "but let's have a snack first."

What planets did the Darin family fly past?

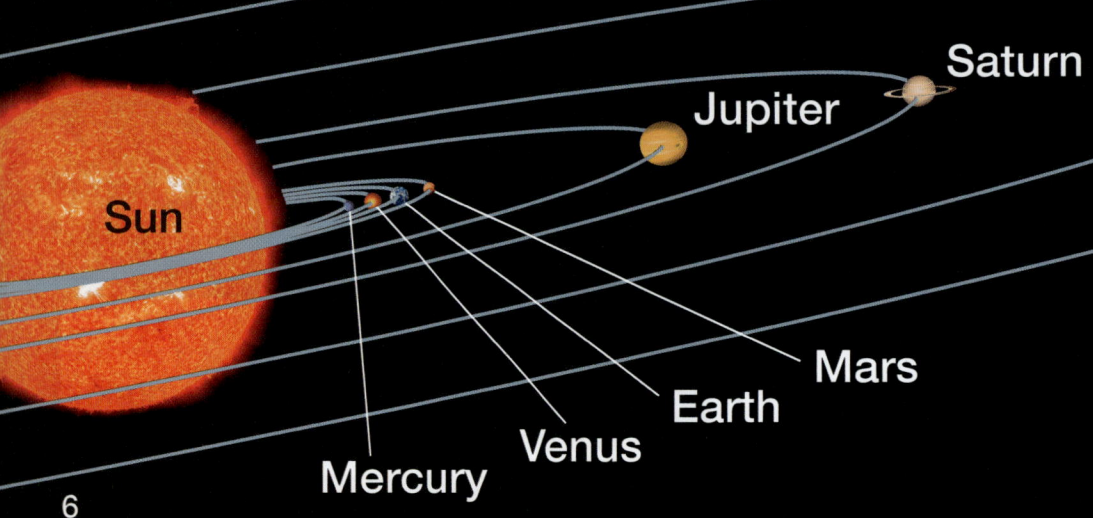

Amazing fact! Jupiter has 95 moons.

Read the words
rock   planet
space   journey

Use the words to draw a picture or make a model. Write about it.

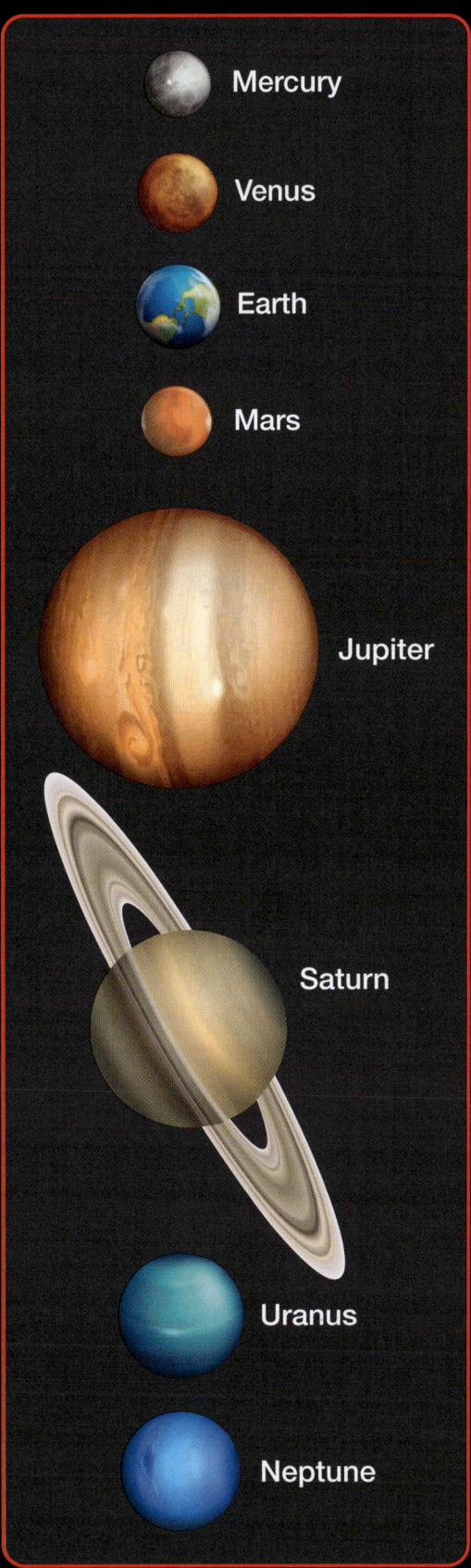

Talking
Which two planets are (a) closest to the sun (b) furthest from the sun?

For you to do
How many planets can you remember?

Day and night

As the Earth spins in space, we get day and night. The sun brings us daylight. In the night it is dark, so we see the moon and stars.

Read and talk about the story.

Flying in the dark

A sparrow was flying back to its nest after a long day looking for food. Every day was the same for the sparrow: fly around looking for food, then back to its nest to sleep.

Along came a bat. The bat flew past the sparrow's nest, and out into the night.

The sparrow slipped out of its nest to follow the bat around the wood. But it was so dark, the sparrow bumped into trees. It couldn't see anything.

So the sparrow flew back to its nest to wait for daytime.

The bat flew through the night. Its special ears use sound to help it fly around. Night-time is an adventure for bats!

What was the difference between the sparrow and the bat?

Read the words

night pole
stars equator
daytime

Use the words to draw a picture or make a model. Write about it.

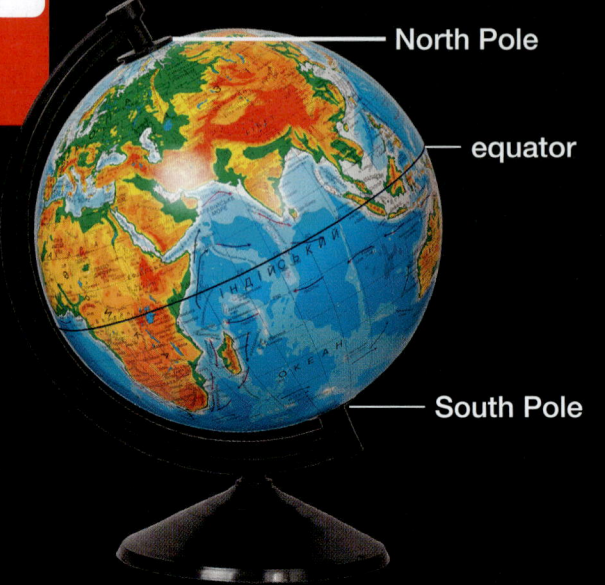

Amazing fact! It never gets dark in the summer at the North Pole.

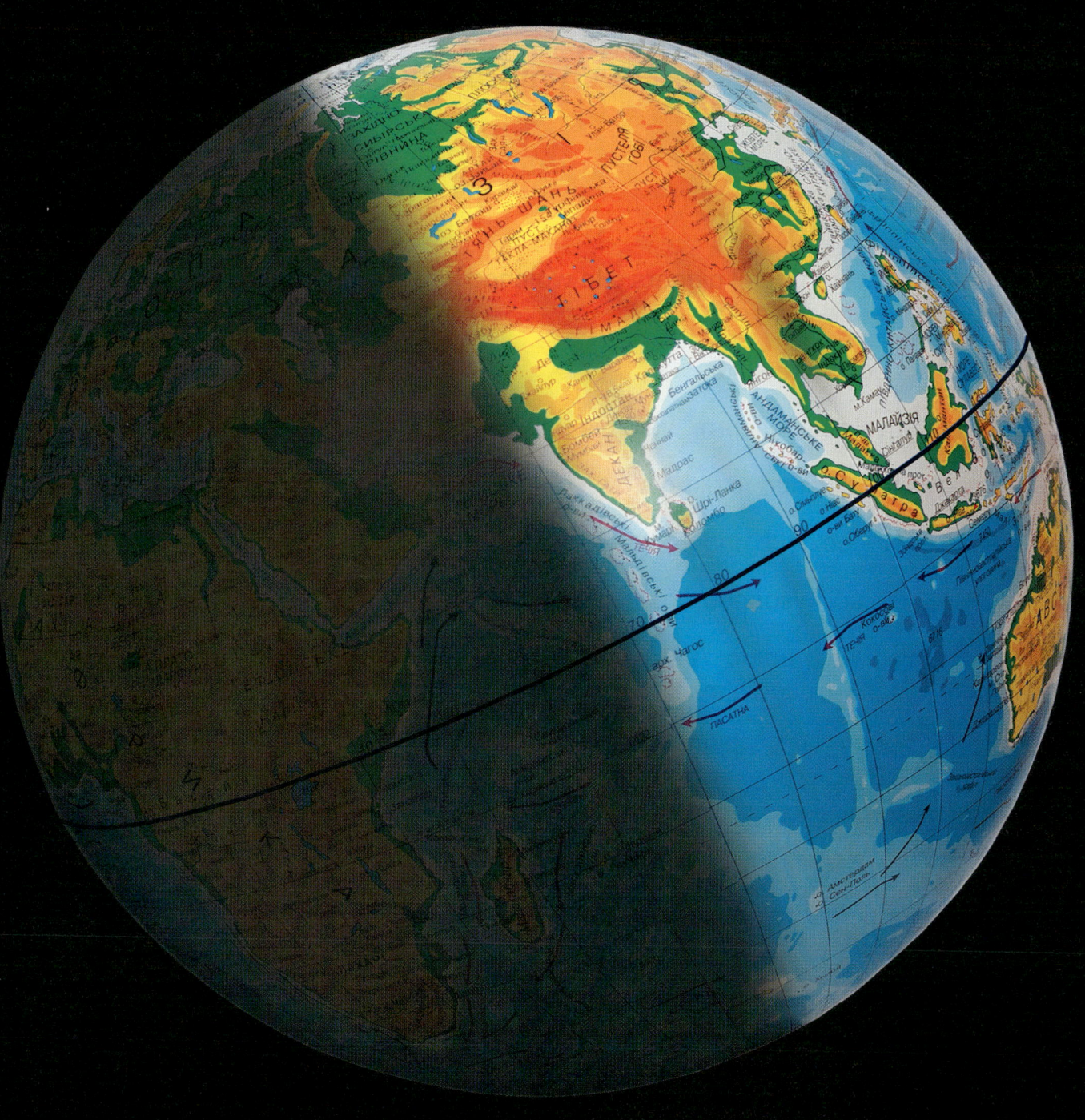

Talking
What do you like about (a) daytime (b) night-time?

For you to do
What happens if you leave lots of lights on at night? Talk about what might happen to the bat and the sparrow.

Land and water

Looking at the Earth from space, astronauts can see land and water. There is much more water than land.

Read and talk about the story.

Learning to dig

Max's family were tunnel diggers. He watched his parents dig new tunnels under the ground each day.

"That looks easy," he said.

"You still have a lot to learn," said his father.

Next day Max decided to dig his own tunnel.

"This is easy," he said, and off he went.

Suddenly the soil disappeared. He couldn't dig.

He couldn't even breathe. A strong hand gripped his arm and pulled him back. His father had rescued him.

"You dug into water," said his father. "Most of the world is covered in water. You can't dig there. You must learn where to dig, and how to swim!"

What did Max learn?

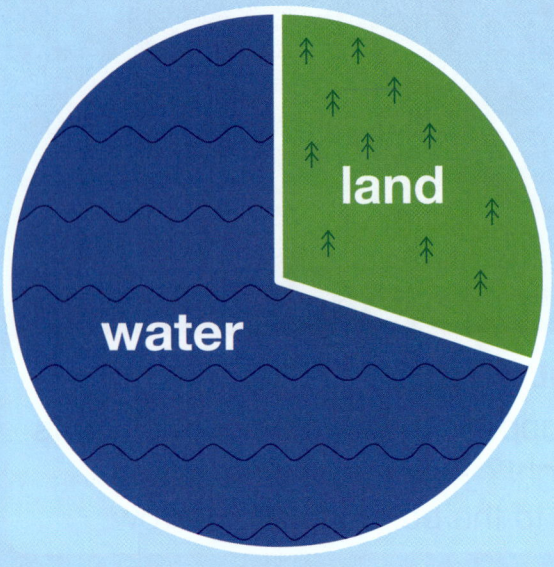

Read the words
water   sea
land   soil

Use the words to draw a picture or make a model. Write about it.

Amazing fact! Nearly all the water in the world is salty.

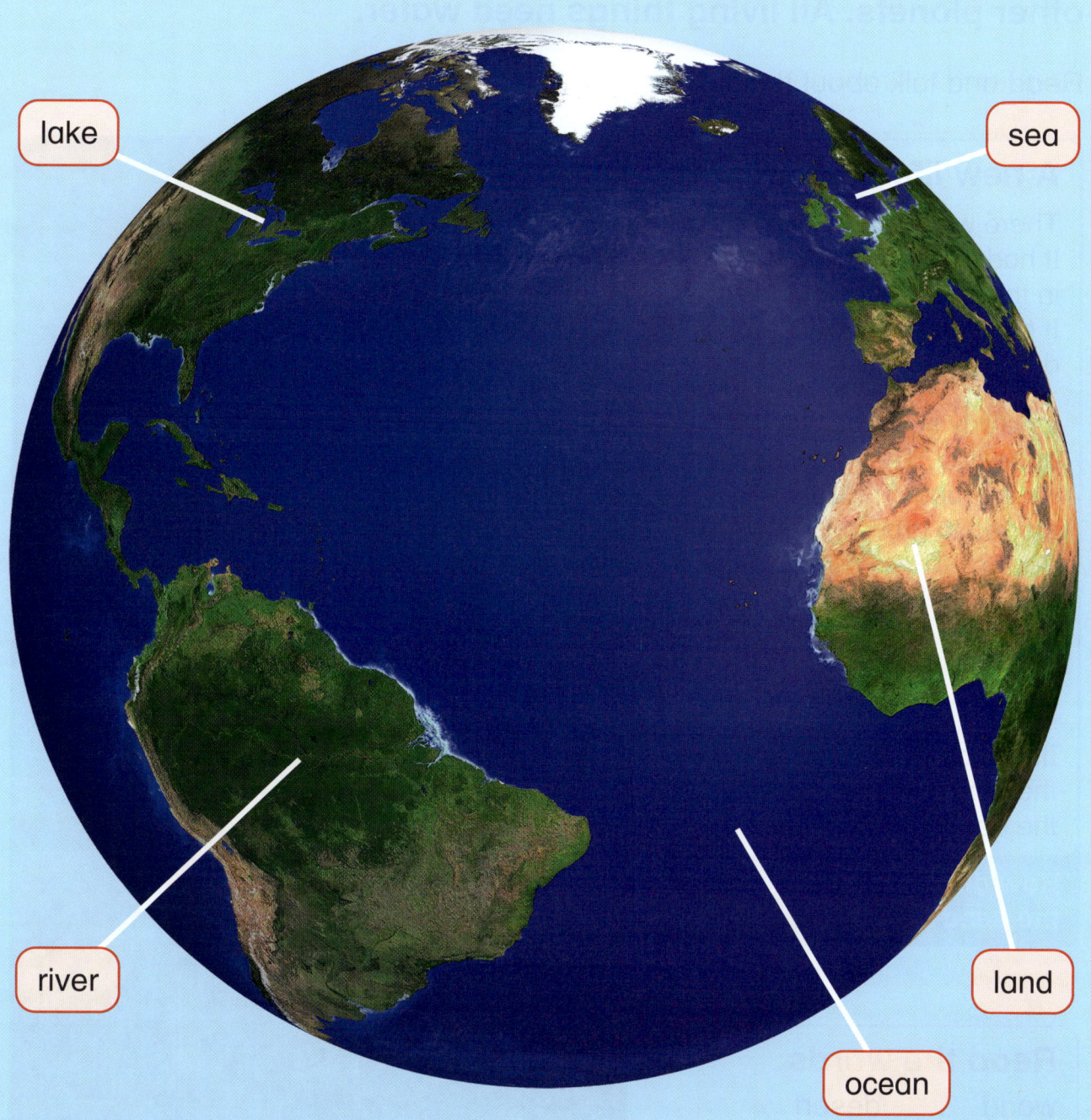

- lake
- sea
- river
- land
- ocean

Talking
Do you think there is more land or water in this image?

For you to do
Take care with how you throw away rubbish. What happens if it ends up in the sea?

A living planet

There is much more water in our world than on other planets. All living things need water.

Read and talk about the story.

A new life

There was once a little seed. It had been lying in the sand in the desert for seven years. It didn't seem to be doing anything.

One day some clouds appeared. They poured rain on the land. The little seed started to grow and grow.

Soon it had roots and branches, then flowers and seeds of its own. The water had brought it to life.

Its seeds would wait for the next rainfall to bring the desert to life again.

How did the rain help the seed?

Read the words

world   desert
seed   cloud
sand

Use the words to draw a picture or make a model. Write about it.

Amazing fact! Some desert areas have no rain for more than 100 years.

beavers in a river

birds and reeds in a pond

fish and coral in the sea

Talking
Why is water important to the plants and creatures in the photographs?

For you to do
Adopt a plant at home or school. See how it changes and water it, if needed.

The shape of the land

Mountains are the highest places on the land. Water flows from mountains to the sea in streams and rivers.

Read and talk about the story.

A journey to the sea

A trout lived in a little pool high in the mountains.

One day, it started to swim to the sea.

Off the trout went, swimming downstream. The water got deeper and the current got stronger. There were cliffs at the side of the valley.

Suddenly the trout was flying in the air … and over a waterfall! A duck quacked close by.

On swam the trout. There was something shiny in the water – a fisher's hook!

The trout dodged the hook and dashed away down the valley, swimming swiftly past an otter.

Finally, the trout reached the sea. What might happen there?

What creatures did the trout meet on its journey to the sea?

Read the words

mountain	current
river	cliff
pool	waterfall
downstream	valley

Use the words to draw a picture or make a model. Write about it.

Amazing fact! The world's longest river is called the Nile. It is over 6000 kilometres long.

Talking
What different words describe the river?

For you to do
Don't use too much toothpaste. It goes down drains into rivers. Why is this a problem?

Volcanoes

There are hot rocks under the ground. They come to the surface in special mountains called volcanoes.

Read and talk about the story.

Strange noises

Azim came running down from his bedroom. "Mum! Mum! I am frightened. I can hear a sound like a volcano," he cried.

"Don't worry," said Mum. "It's only Dad snoring."

Soon Azim came running down again. "Mum! Mum! I am frightened. I can hear a sound like a volcano," he sobbed.

"Don't worry," said Mum. "It's only Dad's tummy rumbling."

Later Azim came running in again. "Mum! Mum! Dad is making a noise like a volcano again."

"No, I'm not," said Dad, waking up. "It must be a real volcano that is about to erupt. Quick everyone, into the car. We must go to grandma's until it is safe to come home again."

Why wasn't it safe for Azim to stay at home?

Read the words
volcano   surface
erupt   core

Use the words to draw a picture or make a model. Write about it.

Amazing fact! Over millions of years, volcanoes have helped to create the air we breathe today.

surface   hot rocks   core

Talking
What happens when a volcano erupts?

For you to do
Where is your nearest hill? Find out if it was once a volcano.

World wonders

There are many special sights in the world. There are wonders on the land, wonders in the sea and wonders in the sky.

Read and talk about the story.

Wonders all around

Azim was at the coast with his grandmother. He was feeling gloomy.

"How can I cheer you up?" said Azim's grandmother.

"I wish I could see something bright and interesting," said Azim.

Suddenly the sun came out from behind a cloud. It shone on a sparkling white iceberg in the sea. "That's better," said Azim, "but it's very cold. I wish I was somewhere warmer."

Azim's grandmother led him into a deep cave full of amazing shapes. "That's better," said Azim, "but it's a bit eerie down here."

Azim and his grandmother went back out onto the beach. Suddenly the sky was filled with whirling colours. "It's the northern lights!" said Azim's grandmother.

"That's better," said Azim. "Now I wish I had some new toys."

"That's enough wishes for one day!" said Azim's grandmother.

What wonders did Azim see?

iceberg

Read the words

coast northern lights
iceberg beach
cave

Use the words to draw a picture or make a model. Write about it.

Amazing fact! Some icebergs contain water that fell to Earth thousands of years ago.

cave

northern lights

Talking
What other world wonders can you think of?

For you to do
What unusual thing could you show a visitor staying with you?

Experiencing the weather

In some countries, the weather does not stay the same for long. There is sun, rain and wind. Sometimes it feels hot. Sometimes it feels cold.

Read and talk about the story.

A nasty surprise

Mika was out for a walk. It was a bright, sunny day. The birds were singing and the bees were buzzing.

Suddenly he felt a spot of rain. "Oh! Never mind," he said. "It will only be a shower." He put up his umbrella.

Soon the wind got up. The gentle breeze turned into a howling gale.

It blew into Mika's umbrella. "Help!" he cried. "I'm flying in the air."

The wind dropped and so did Mika. Into a pond.

"An umbrella was meant to keep me dry," he moaned. "The weather just isn't fair."

How was Mika caught out by the weather?

If it rains, you might use an umbrella.

Read the words

weather   shower
rain   gale
wind   snow

Use the words to draw a picture or make a model. Write about it.

Amazing fact! It almost never rains in Antarctica.

If it is windy, you can fly a kite.

If it snows, you can make a snowman.

Talking
What different types of weather have you experienced? What is your favourite?

For you to do
Where is the best part of the playground to go when it is very windy?

Different types of weather

We use different words to talk about the different types of weather. Sometimes we show what the weather is like using pictures called symbols.

Read and talk about the story.

Mika's holiday

Mika was looking forward to going on holiday. "I need a rest," he said.

He set off in his car in bright sunshine. The car's paint shone and the engine hummed in the warmth.

Before long, some clouds appeared. "It's nothing," Mika said. "It'll soon change."

It did! It began to rain. "It's nothing," he said, and closed the windows. "It'll soon change."

It did! There was thunder and lightning. Water sprayed over the car. Mika put on the headlights. "It's nothing," he said. "It'll soon change."

It did! He ran into fog! "This is awful," he complained. "I'm stopping altogether. It'll never change."

Suddenly, the sun came out.

What different types of weather did Mika experience?

Warm sunshine

Cloud and rain

Read the words

symbols	rain	thunder
warm	shower	lightning
sun	fog	
cloud	mist	

Use the words to draw a picture or make a model. Write about it.

Amazing fact! The water in a rain drop never ever disappears.

Sun and cloud

Sun and showers

Fog and mist

Thunder and lightning

Talking
What weather words can you think of?

For you to do
Always wear the right clothes for the weather. Why does this matter?

Extreme weather

Sometimes the weather is wild and exciting. It can be very rainy, cold, windy or hot. Tornadoes, hurricanes, floods and sandstorms are very powerful.

Read and talk about the story.

A day off school!

Priya walked slowly home from school. "I wish I could have a day off," she grumbled.

That night, she was woken up by a crash and a rumble. She looked out of the window. The sky was full of dark clouds. The trees were bent over by the wind.

Suddenly, a great flash lit up the sky. It began to rain and rain and rain. In the morning, there was water everywhere. The fields were flooded.

"Hurray! No school today," she thought. She went back to bed.

There was a knock on the door. "Come on out," shouted a voice. It was Priya's teacher. She was collecting everyone in her boat.

Poor Priya!

Do you think Priya was frightened by the weather?

Read the words
field　　flood
tornado　sandstorm
hurricane

Use the words to draw a picture or make a model. Write about it.

Amazing fact! There are millions of thunderstorms around the world each year.

tornado

sandstorm

hurricane

flood

Talking
What weather might close your school?

For you to do
Imagine what it might be like to be in these photographs. Talk to a partner about your ideas.

The seasons

In some countries, for example, the United Kingdom, there is a pattern of seasons during the year. In winter, people stay inside to keep warm. In summer, they like to be outside.

Read and talk about the story.

The lazy rabbit

A little rabbit was born in the spring. Outside its burrow it found sweet, juicy grass. By summer it was big and strong.

While the other rabbits spent lots of time repairing their burrow, this rabbit just slept outside. It was sunny and there was plenty of food.

In autumn, it rained. The rabbit had to take carrots from a garden and sleep under some old hay.

Soon it began to snow. The rabbit was cold without a burrow to keep warm.

Maybe the rabbit will be allowed back into the burrow if it works really hard next year.

Do you think the rabbit will work hard?

Why did the rabbit enjoy the spring and summer more than the autumn and winter?

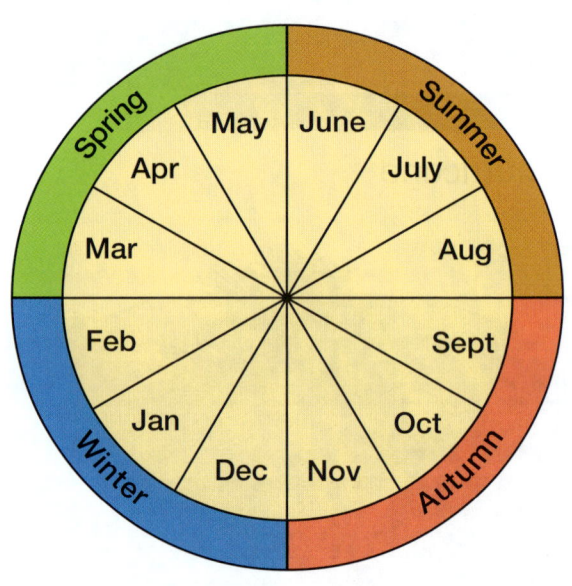

Read the words

seasons   summer
pattern   autumn
year   winter
spring

Use the words to draw a picture or make a model. Write about it.

Amazing fact! In the UK there can sometimes be winter weather even in the middle of summer.

Spring

Summer

Autumn

Winter

Talking
Is there a pattern of seasons where you live?

For you to do
What do you like doing at different times of the year?

Going round the sun

The seasons change as the Earth goes round the sun. The sun is high in the sky in summer and low in winter.

Read and talk about the story.

The children can't agree

"I don't like winter," said Baby Oscar. "It's so cold."

"Summer is best," said Theo. "Hot sun for me!"

"Why do seasons change anyway?" asked Ava.

"I think it's the trees," said Baby Oscar. "When they are tired, it's winter. When they wake up, it's summer."

"I think it's the sun," said Theo. "When its fires are bright, it's summer. When they die down, it's winter."

"That doesn't seem right," said Ava. "It's to do with the Earth going round the sun."

They all began to argue.

What do you think?

Who had the best answer?

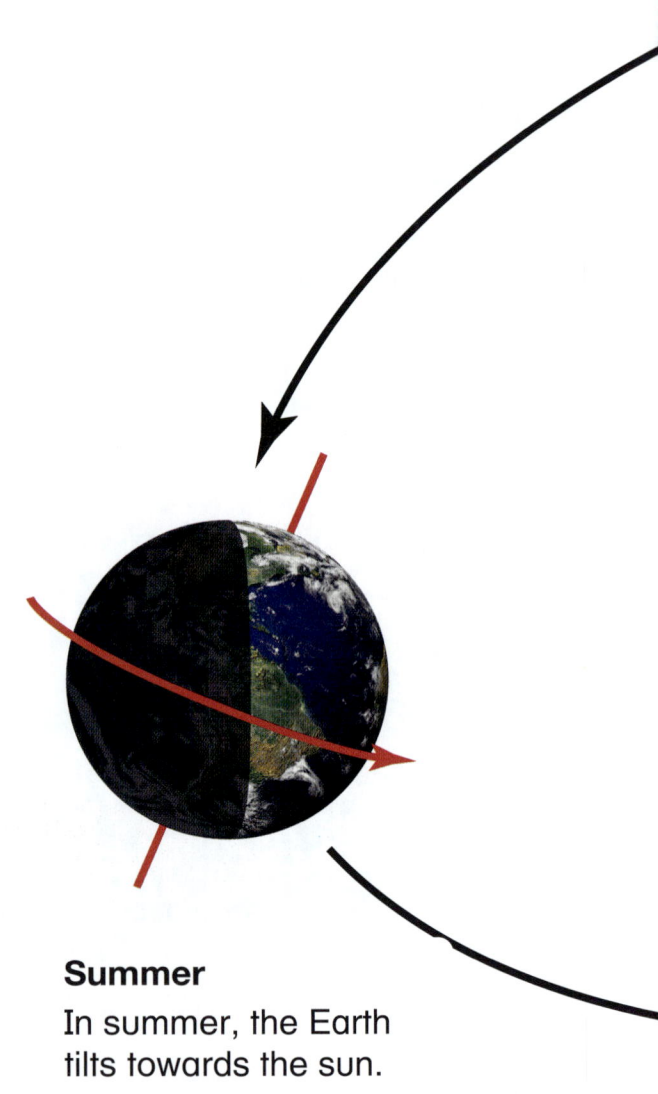

Summer
In summer, the Earth tilts towards the sun.

Read the words

high	tilt	away
low	towards	

Use the words to draw a picture or make a model. Write about it.

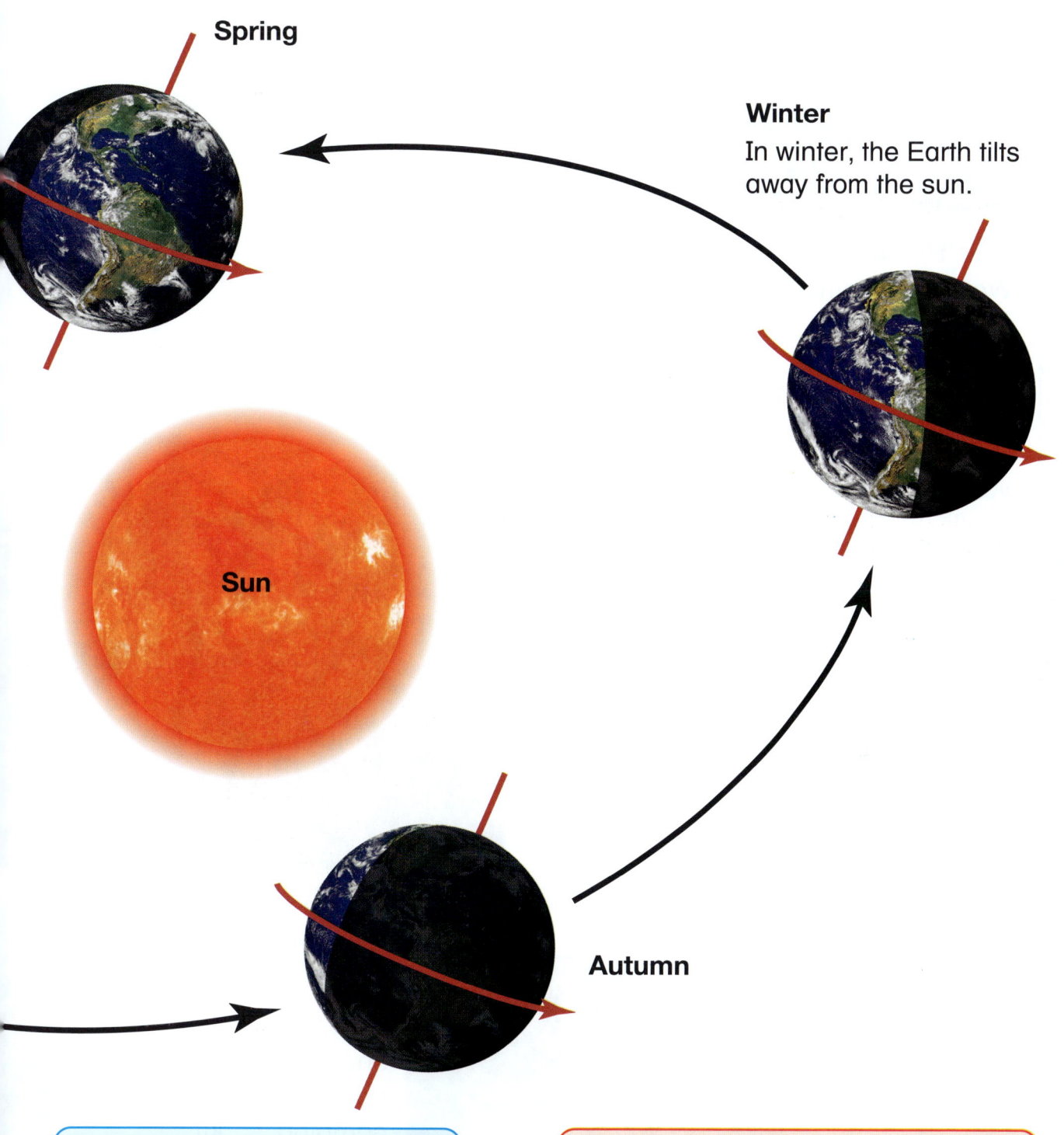

Amazing fact! Mercury takes just 88 days to go round the sun. Neptune takes 165 years.

Winter
In winter, the Earth tilts away from the sun.

Talking
In which season does the Earth (a) tilt towards the sun (b) tilt away from the sun?

For you to do
Give garden birds food and water when it is very cold or hot. Why do we need to help them?

Shelter

Homes give us shelter from the weather. They are where we feel warm and safe. Some people make homes that they can move from place to place.

Read and talk about the story.

Camping in the garden

"We want to camp in the garden," said Toya and Efe.

"No," said their mum. "You won't like it."

"We will! We will!" answered the children.

"Well I did warn you," replied Mum.

That afternoon, it was warm and sunny. The children put up their tent. They put sleeping bags, food and drink inside.

"It's our little home," they said.

In the night, the wind began to blow. The trees rustled. An owl hooted. "I'm frightened," said Toya.

"It's a bit cold," said Efe.

In the morning, Mum found them indoors, safely tucked up in bed.

Why did Toya and Efe think their tent had become a little home?

Read the words

shelter	tent
houses	item
camping	kit

Use the words to draw a picture or make a model. Write about it.

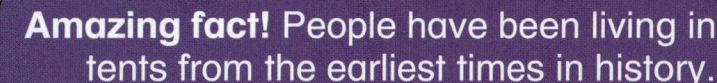

Amazing fact! People have been living in tents from the earliest times in history.

Camping kit: cooking pots, tent, sleeping bag, water containers, table and chairs, games and toys

Talking
How can the items in the camping kit make a tent a home?

For you to do
What jobs could you do at home to help your family?

Houses around the world

People build houses in different ways. They can be made of brick, wood, stone or concrete. All houses give us shelter.

Read and talk about the story.

City and country

Ewan lived in the country. His cousin Isla came from the city for a holiday.

"You will like it here," said Ewan. "It's peaceful and quiet."

They walked through the village. They passed little cottages and old wooden houses. At the end was a little caravan park with holiday homes. That was all.

After a week Isla said, "It's very quiet. Come to the city with me."

Off they went. Isla took Ewan along her street. There were single houses, houses joined together and great blocks of flats. People and cars rushed everywhere. They jumped back as a police car with flashing lights sped by.

"I'm going back to the country," said Ewan. "The city is too busy and noisy for me!"

What types of homes did Ewan and Isla see?

Read the words

brick	concrete
wood	country
stone	city

Use the words to draw a picture or make a model. Write about it.

Amazing fact! The world's tallest buildings are as high as many mountains.

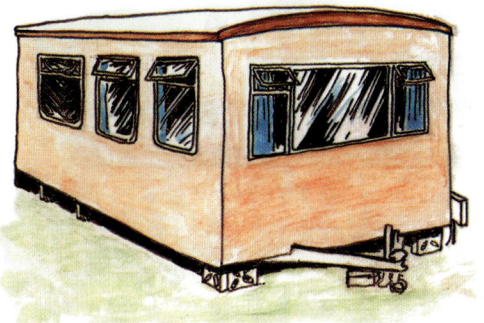

Talking
Are the homes in the drawings all built of the same materials?

For you to do
Say the address of where you live. What does each line mean?

Living in a village

A village is a place where houses are grouped together. There are also schools, shops and other places which help people in their daily lives.

Read and talk about the story.

Ewan loses his hat

Ewan was feeling sad. He could not find his hat. He went from one end of the village to the other, asking everyone, "Did I leave my hat here?"

He asked in the shop, but they said, "No."

He asked at the garage, but they said, "No."

He asked at the church, but they said, "No."

He asked the people at the big house and in the cottages, but they all said, "No."

The last place in the village was the school.

"Have you seen my hat?" he asked.

"Yes," they said, "it's on your head!"

What places did Ewan visit?

shop

cottages

Amazing fact! Nearly every village in the United Kingdom is hundreds of years old. A village in the UK always has a church.

Read the words

village   shop   cottage
school   garage   pond

Use the words to draw a picture or make a model. Write about it.

church

pond

school

Talking
Why is a village more than just a group of houses?

For you to do
How do people who deliver letters and parcels know where you live? Does your home have a number that is easy to read?

Exploring local streets

There are lots of things to see in the streets around you. These are clues to the way we live and the things we need.

Read and talk about the story.

Read the words

street   lamp post
rubbish bin   drain

Use the words to draw a picture or make a model. Write about it.

The fancy dress party

Isla was going to a fancy dress party dressed as a pirate. She had a pirate's hat, a red shirt, long boots and a cutlass made from cardboard and silver paper.

She ran quickly down the street. She bumped into a rubbish bin and tore her trousers.

She brushed against a newly painted bus stop and spoiled her shirt.

Her hat was knocked off by a lamp post and squashed by a bus.

Her boot caught in a drain and the heel broke. A bird flew away with her cutlass.

Isla was in a state when she arrived at the party. But she still won a prize. For being dressed as a scarecrow!

What was in Isla's way as she rushed down the street?

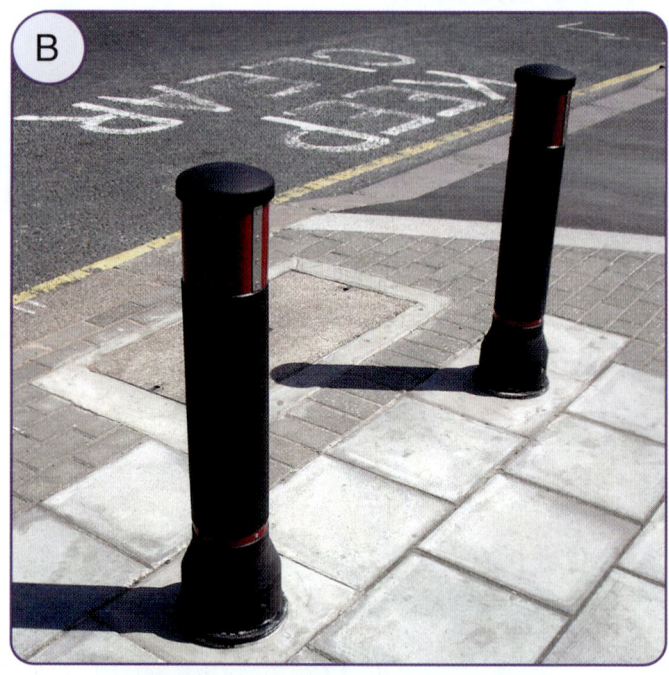

Amazing fact! The first street lights were in China 2500 years ago.

D

S

P

L

Talking
What should you not put down drains? Where does a drain lead to?

For you to do
Use the letters in the circles to help you work out what each photograph shows: postbox, lamp post, sign, park, drain, bollard.

Under your feet

There are lots of pipes and wires under the pavement.

Read and talk about the story.

> **Read the words**
> pipes   pavement   electricity
> wires   gas   ground
>
> **Use the words to draw a picture or make a model. Write about it.**

Ewan's new house

Ewan liked his new house.
Every morning he switched on the light, filled the saucepan and cooked an egg for breakfast.
One day he got up and switched on the light, but it did not come on.
He called for help. "No electricity," he said.
Soon there was a knock at the door. It was Mr Diggery.
"What can you do?" asked Ewan.
"Dig up the road," said Mr Diggery.
He found a broken wire.
The next day Ewan got up, switched on the light and tried to fill the saucepan, but there was no water.
He called for help. Soon Mr Diggery came.
"What can you do?" asked Ewan.
"Dig up the road," said Mr Diggery.
He found a broken water pipe.
The next day Ewan got up, switched on the light, filled the saucepan, but there was no gas.
He called for help. Mr Diggery came.
"What can you do?" asked Ewan.
"Dig up the road," said Mr Diggery.
The gas pipe was blocked.
The next day Ewan got up and had his breakfast.
"I will go for a walk," he said.
He opened the front door and fell into a big hole.
"Help! Help!" he cried.
"Today I can dig YOU up," laughed Mr Diggery.

What did Ewan find out about his house?

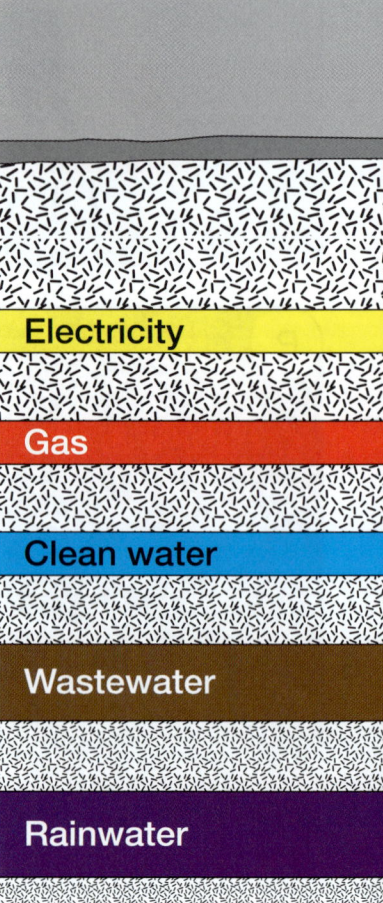

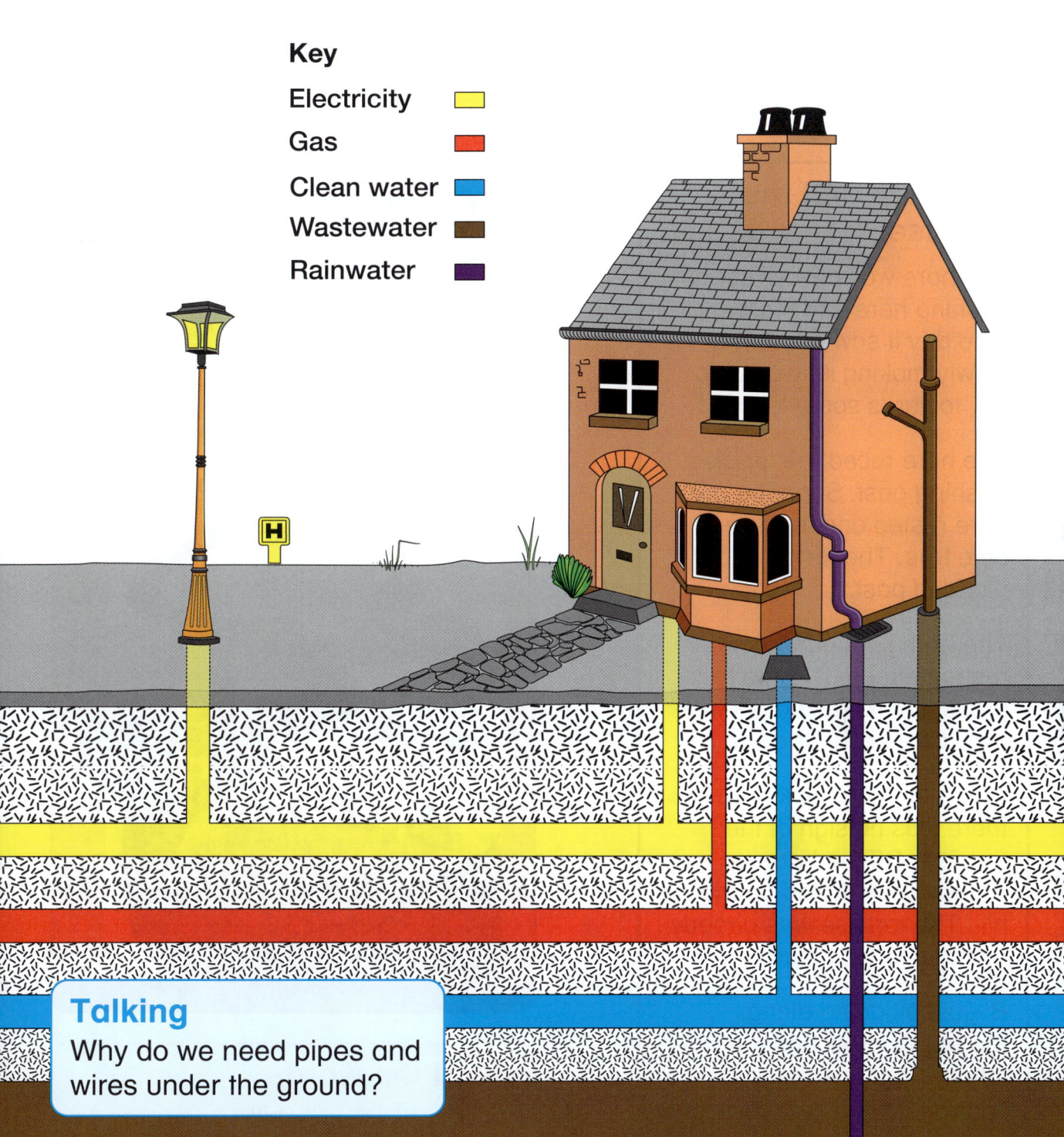

Amazing fact! The amount of water we use in a week weighs about the same as a small car.

Key
- Electricity
- Gas
- Clean water
- Wastewater
- Rainwater

Talking
Why do we need pipes and wires under the ground?

For you to do
Turn off the light when you leave a room at night. Why does this matter?

Maps and stories

Picture maps can show us the places in songs and stories. They make it easier for us to see what happens.

Read and talk about the story.

The hare and the tortoise

The hare was always speedy, dashing here and there. One day it saw a tortoise slowly making its way up a hill, towards some lettuces.

The hare raced the tortoise, dashing past. Soon the hare rested and fell asleep by a tree. The tortoise plodded past.

The hare jumped up and rushed ahead. It came to a grassy bank, rested again and dozed off in the sun.

When the hare woke up, there was no sign of the tortoise. Off the hare ran. At last it scrambled up the hill. The tortoise was already eating a lettuce leaf!

By just plodding along, the tortoise reached the lettuces first.

What places did the hare and the tortoise pass in their race?

Read the words

map	hill
place	bank

Use the words to draw a picture or make a model. Write about it.

Amazing fact! People started making maps before they learnt to read and write.

examples of journey maps

a picture map of a fairy tale kingdom

Talking
What other stories and songs do you know which are about places?

For you to do
Ask an older person for stories and memories of where you live. Draw a story map.

Treasure island

Maps show us where places are and how they are linked together. They help to stop us getting lost.

Read and talk about the story.

A lucky find

Toya and Efe were excited. Their father had bought a metal detector.

Sam, the boy next door, was jealous. He made a pretend treasure map to trick them. "My grandad had this old map," he said, laughing to himself. "See if you can find treasure with your detector."

Toya and Efe excitedly took the map. They started off, going west from the town to the windmill. Nothing there. They went south to the lighthouse. Nothing there. They went east to the castle.

They were going to give up. Suddenly the metal detector beeped. They dug a hole and found an old Roman coin.

"We did find treasure," they told Sam when they returned. "It's going to the museum." Sam wasn't laughing now.

What directions did Toya and Efe go in?

Read the words

town   directions
lighthouse   island
castle   grid
museum

Use the words to draw a picture or make a model. Write about it.

Amazing fact! There are around 25 000 islands in the Pacific Ocean.

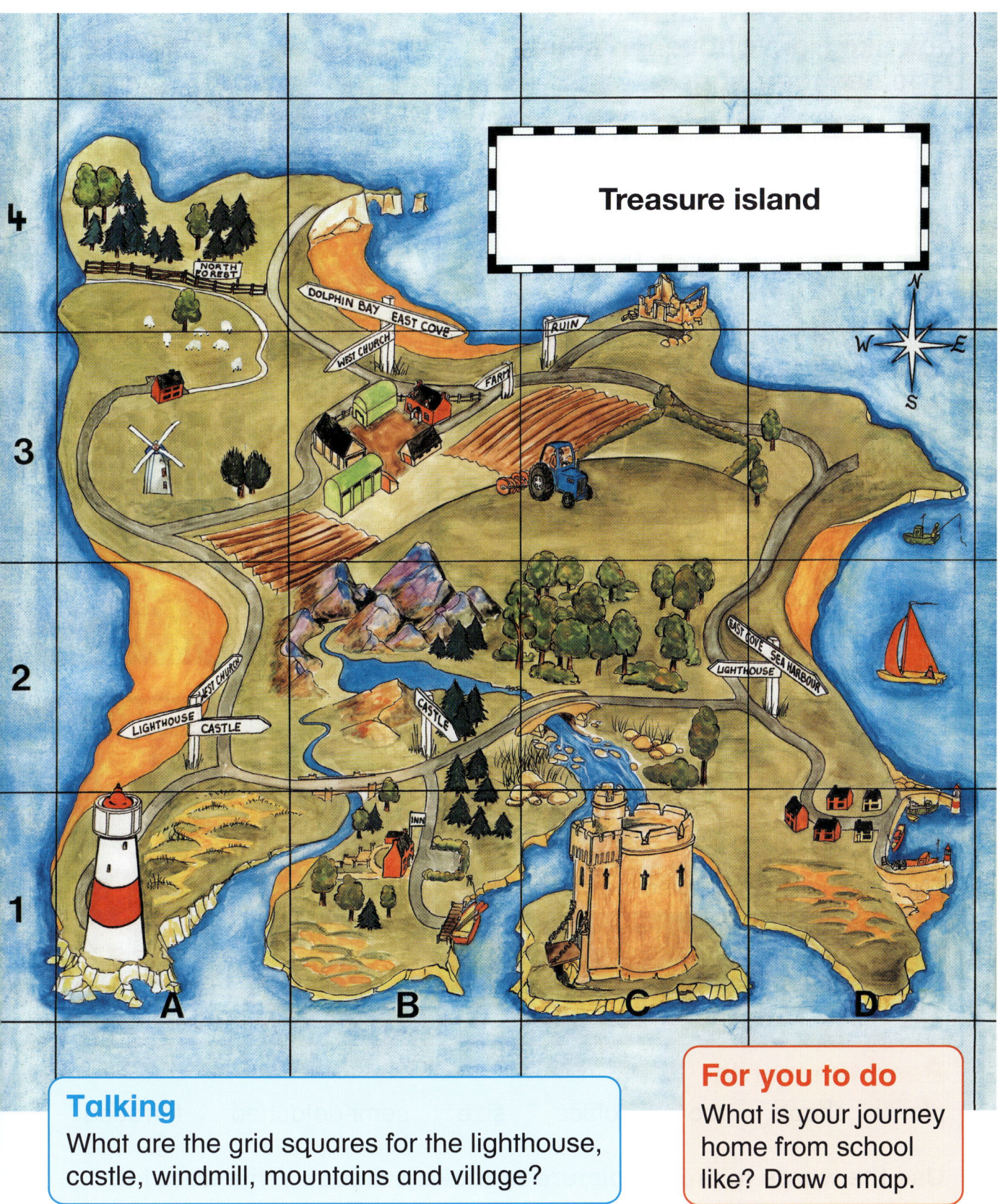

Talking
What are the grid squares for the lighthouse, castle, windmill, mountains and village?

For you to do
What is your journey home from school like? Draw a map.

43

Different plans

Plans show us what places look like from above. This helps us to see their shape.

Read and talk about the story.

A new school

The children were excited. Their new school was opening.

Class 1, the youngest, went into their classroom first. It was just right.

Class 2 went in next. The children could only just fit in their room.

Class 3, the eldest, were last. "This is hopeless," they complained. "There is only space for half of us."

The head teacher called the children into the hall. It was so small, Class 1 filled it up!

"My office is too big," said the head teacher. "So is mine," said the secretary.

The builder came. "I expect you are pleased with your new school," he said. "It was quick to build because I decided to make all rooms the same size."

"What a bad idea!" shouted the children.

Why can't all rooms be the same size and shape?

Read the words

plan   shape   room   office   size   semi-detached   terraced

Use the words to draw a picture or make a model. Write about it.

Amazing fact! The buildings on Palm Island in Dubai are arranged in the shape of a palm tree.

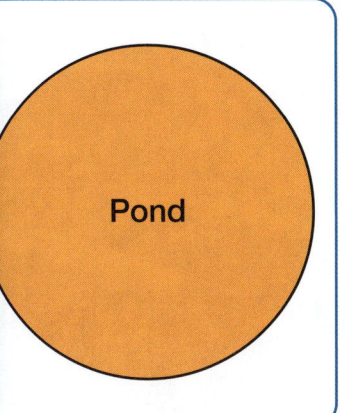
Pond

Roundabout

School

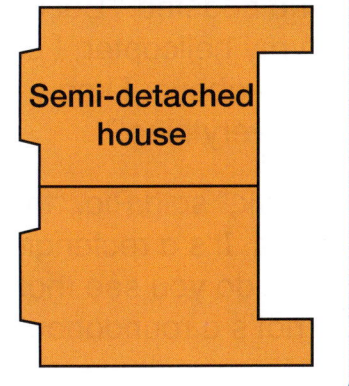
Semi-detached house

Shop

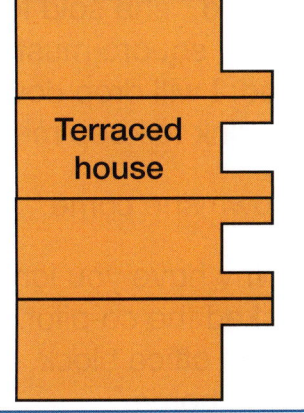
Terraced house

Talking
What is the shape of some of the buildings and places that you know?

For you to do
What shape is your classroom? Make a plan of it.

45

The view from above

Plans show what places look like from above. The higher we go in the sky, the smaller places seem to be.

Read and talk about the story.

Pia the proud pilot

Pia the pilot was feeling very proud.

"I know what everything looks like," she boasted. "Because I fly so high in my helicopter, I can see things from above. I know the shapes of everything."

"Look," said Pia, "there is a football pitch. It's a rectangle shape. And do you see that circle? That's a roundabout."

Suddenly Pia spotted a square.

"Good," she said to herself. "That square must be my landing pad. I will drop down and have a drink to keep me going."

Down she came.

"Why have you landed here?" asked the co-pilot. "We're on top of an office block, many kilometres from home."

"Oh no!" groaned Pia. "It looked just the right shape."

Why did Pia get confused?

Read the words
view   rectangle
above   photograph

Use the words to draw a picture or make a model. Write about it.

Amazing fact! Astronauts can see the pyramids in Egypt from space.

Talking
What different shapes can you see in the photograph?

For you to do
Apart from school, what other places can you think of near where you live?

Countries and capitals in the United Kingdom

There are four countries in the United Kingdom. England is the largest. Northern Ireland is the smallest. The other two are Scotland and Wales. The capital cities of the United Kingdom are London (England), Belfast (Northern Ireland), Edinburgh (Scotland) and Cardiff (Wales).

Read and talk about the story.

Arjun tidies his room

Arjun was reading his book.

"Please tidy up," said Dad.

Arjun looked at the messy room and grumbled. "Oh dear! I wish I wasn't here." He turned back to his book … and imagined he was inside the story. Arjun was in a fir tree on a mountain.

"Where's this?" he asked. "Scotland," answered a forester. "It's cold," shivered Arjun. "Take me away."

He flicked to the next chapter. Now he was on rocks by a stormy sea.

"Where's this?" he asked. "Northern Ireland on the Giant's Causeway," answered a walker.

"Help!" cried Arjun. "I don't like giants."

He flicked to another chapter. Now he was on a grassy hillside surrounded by sheep.

"I'm hungry," he said. "Where's this?"

"Wales," answered a farmer. "Have some grass."

"Back to England," wished Arjun.

He closed his book. He was home.

"Finished tidying?" asked Dad.

"Not quite," said Alfie.

Which countries does Arjun imagine he visits?

Edinburgh Castle, Edinburgh, Scotland

Houses of Parliament, London, England

Amazing fact! Great Britain (England, Wales and Scotland) is the ninth largest island in the world.

Read the words

country	capital city
England	London
Northern Ireland	Belfast
Scotland	Edinburgh
Wales	Cardiff

Use the words to draw a picture or make a model. Write about it.

Talking
What things do you know about each country in the United Kingdom?

For you to do
Which country in the United Kingdom would you like to visit? Find out more about it.

Mountains, rivers and seas in the United Kingdom

There are groups of mountains in the north and west of the United Kingdom. Rivers flow through the flatter land in the south and east. Rivers flow into seas and oceans. There are three seas around the United Kingdom.

Read and talk about the story.

Yr Wyddfa (Snowdon)

River Severn

Norfolk Broads

Time for a story

Mrs Jones, the teacher, decided to tell the children a story. "I'm going to tell you about a special place in the United Kingdom," she said.

"Would you like to hear about the little train going up Yr Wyddfa in a snowstorm?"

"No! No!" chorused the children.

"Would you like to hear about the journey of a baby eel swimming up the River Severn?"

"No! No!" cried the children.

"What about a holiday adventure on a boat in the Norfolk Broads?"

"No! No!" called the children.

"Well let's do some maths instead," said Mrs Jones.

"No! No!" roared the children.

"Yes! Yes!" said Mrs Jones.

What three stories did Mrs Jones offer to tell the class?

Read the words
Yr Wyddfa (Snowdon)
River Severn
Norfolk Broads

Use the words to draw a picture or make a model. Write about it.

Amazing fact! Some mountains in the United Kingdom were once active volcanoes.

Talking
What are the names of the mountains, rivers and seas shown on the map?

For you to do
Imagine what the place where you live looked like before there were buildings.

Living in the Arctic

The Arctic surrounds the North Pole. It is very cold and snowy. There are long, dark winters. Animals struggle to survive in the Arctic.

Read and talk about the story.

Surprise for Nimtok

It had been snowing all night. Nimtok looked out of his bedroom window to a see a blanket of white covering the land.

"No school today," said his father, "we can't travel far from home."

"Oh dear!" thought Nimtok. "I can't even play with my friends."

After breakfast he put on warm clothes, a furry hat and boots, and wandered into the snow.

He rolled a snowball, then began to make a snowman, but it wasn't much fun.

"I know," he said, "I'll make some big models of my favourite animals."

He began with the walrus because that was an easy shape. The Arctic fox was harder but it wasn't too big. The reindeer was the hardest but he did his best.

He called to his father to come and see. "These are very good," said his Dad. "I like the polar bear best."

"I didn't make a bear!" cried Nimtok. "It must be real."

"Run! Run!" they both shouted.

What animals did Nimtok make with snow?

Read the words
Arctic   walrus   polar bear
North Pole reindeer

Use the words to draw a picture or make a model. Write about it.

Amazing fact! There is water, not land, under the ice at the North Pole.

polar bear

reindeer

walrus

Talking
Would you like to visit the Arctic?

For you to do
If it is cold, wear warm clothes. Why is it better not to use too much heating in winter?

Living in the rainforest

The rainforest lies on the equator. It is hot and wet. Plants grow quickly. There are many different animals in the rainforest.

Read and talk about the story.

Isla's forest walk

Isla was on holiday near the Amazon River.

One morning she decided to go for a walk to see what the rainforest was like.

As she set off, her cousin Ewan followed behind her.

Soon the path she was following became blocked by vegetation. Tall trees began to cut out the light, and bushes and creepers brushed her arms and legs.

"I can hear so many animals," she said, "but I can't see anything moving."

Suddenly a bird fluttered high in a tree. "Oh! A magpie," said Isla.

"No it's not," said Ewan. "That's a parrot."

Further on, a tail disappeared around a corner. "Oh! A tiger," cried Isla.

"No it's not," said Ewan. "That's a jaguar."

Isla stopped to wipe her brow. It was very hot and sticky. There was a flash of colour. "A butterfly!" she gasped.

"No it's not," said Ewan. "That's a hummingbird."

Soon Isla decided to turn back. An animal dashed across the path. "Oh! A rabbit," she whispered.

"No it's not," said Ewan. "That's an anteater."

Back in town Isla took out some money from her purse.

"That's for me," said Ewan.

"No it's not," laughed Isla. "It's for a cup of tea."

What different creatures did Isla and Ewan see?

Amazing fact! The rainforest is home to half of the world's plant and animal species.

Read the words

rainforest   jaguar
vegetation   hummingbird
parrot   anteater
butterfly

Use the words to draw a picture or make a model. Write about it.

Talking
What are all the words you know which are about the rainforest?

For you to do
Ask an adult about how trees grow.

Living in the desert

Most deserts are very hot and dry. Plants and creatures have to protect themselves from the heat.

Read and talk about the story.

lizard

Stranded in the desert

Tami lived in Australia. She decided to explore the desert sights.

Her friends warned, "If your car breaks down, stay in the shade and drink plenty of water."

The desert went on for thousands of kilometres. It was bigger than Tami ever imagined. She drove on and on. She ran out of petrol. "What shall I do?" she sobbed.

A ranger stopped by. "Are you in trouble?" he asked.

"Yes," said Tami. "Please find help!"

Off went the ranger. Later Tami heard a helicopter.

The pilot rushed over. "Here is the water you wanted," he said.

"I needed petrol," cried Tami.

"Never mind," said the pilot. "You can ride back with me."

Why was Tami in danger when her car ran out of petrol?

kangaroo

Read the words

shade   lizard
kangaroo   eucalyptus

Use the words to draw a picture or make a model. Write about it.

Amazing fact! Deserts cover one third of the land on Earth.

Uluru/Ayers Rock

desert road

water bottles

eucalyptus trees

helicopter

Talking
Can you retell the story of Tami's adventure? Add information from the photographs.

For you to do
Remember not to waste food. What makes it difficult to grow food in the desert?

Animals around the world

It has taken millions of years for different plants and creatures to develop. We need to remember that we share the world with them.

Polar bear
Greenland

Eagle
Rocky Mountains

Butterfly
Amazon rainforest

King penguin
Antarctica

> **Read the words**
> Greenland   India
> Amazon   Africa
> China
>
> Use the words to draw a picture or make a model. Write about it.

Amazing fact! The blue whale is the largest creature on Earth. It is as heavy as 30 elephants.

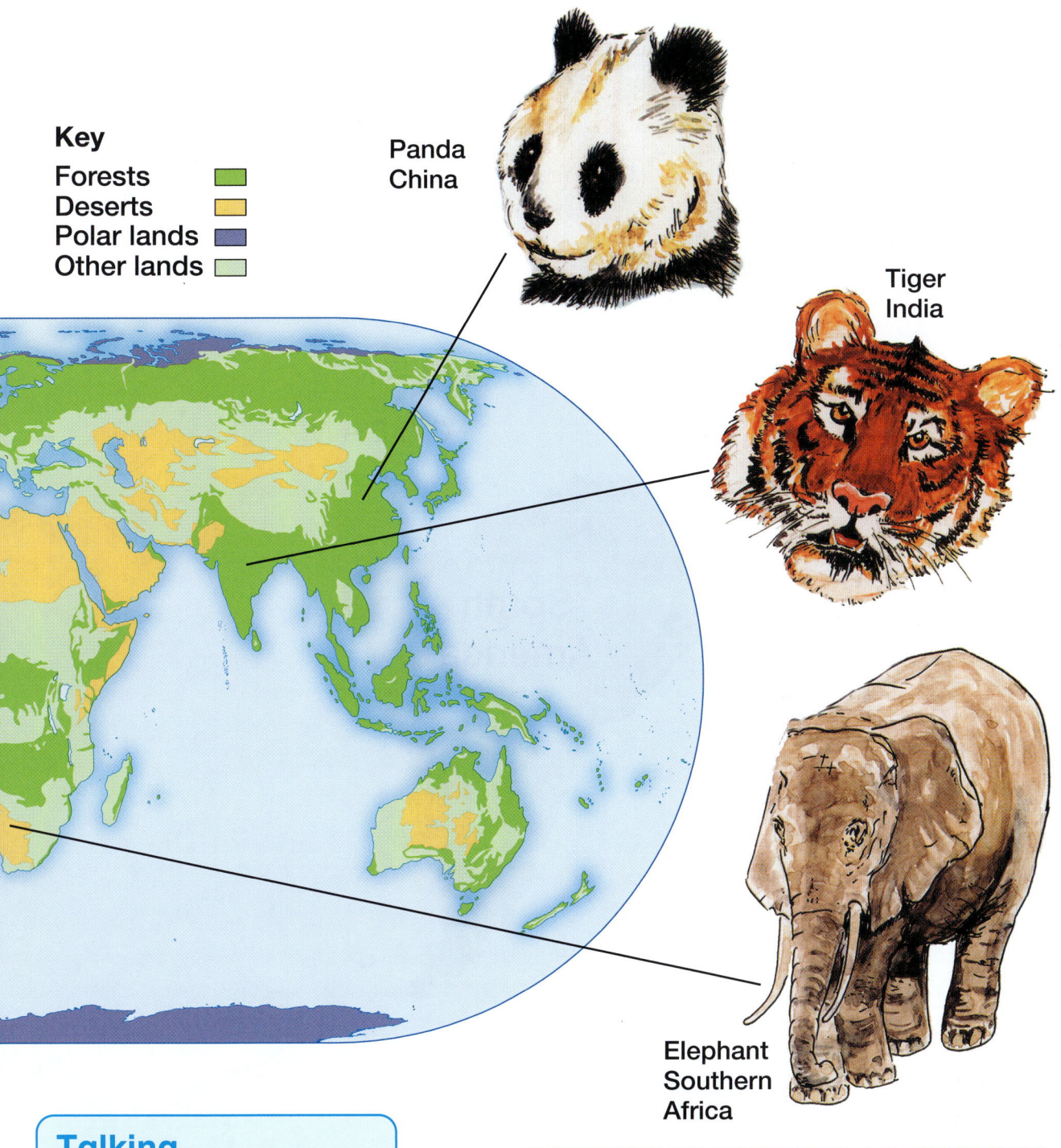

Key
Forests
Deserts
Polar lands
Other lands

Panda
China

Tiger
India

Elephant
Southern Africa

Talking
Which of the creatures live in (a) the rainforest, (b) the desert and (c) polar lands?

For you to do
Ask a teacher about having a bush in school that butterflies like. What other things might you do?

World continents and oceans

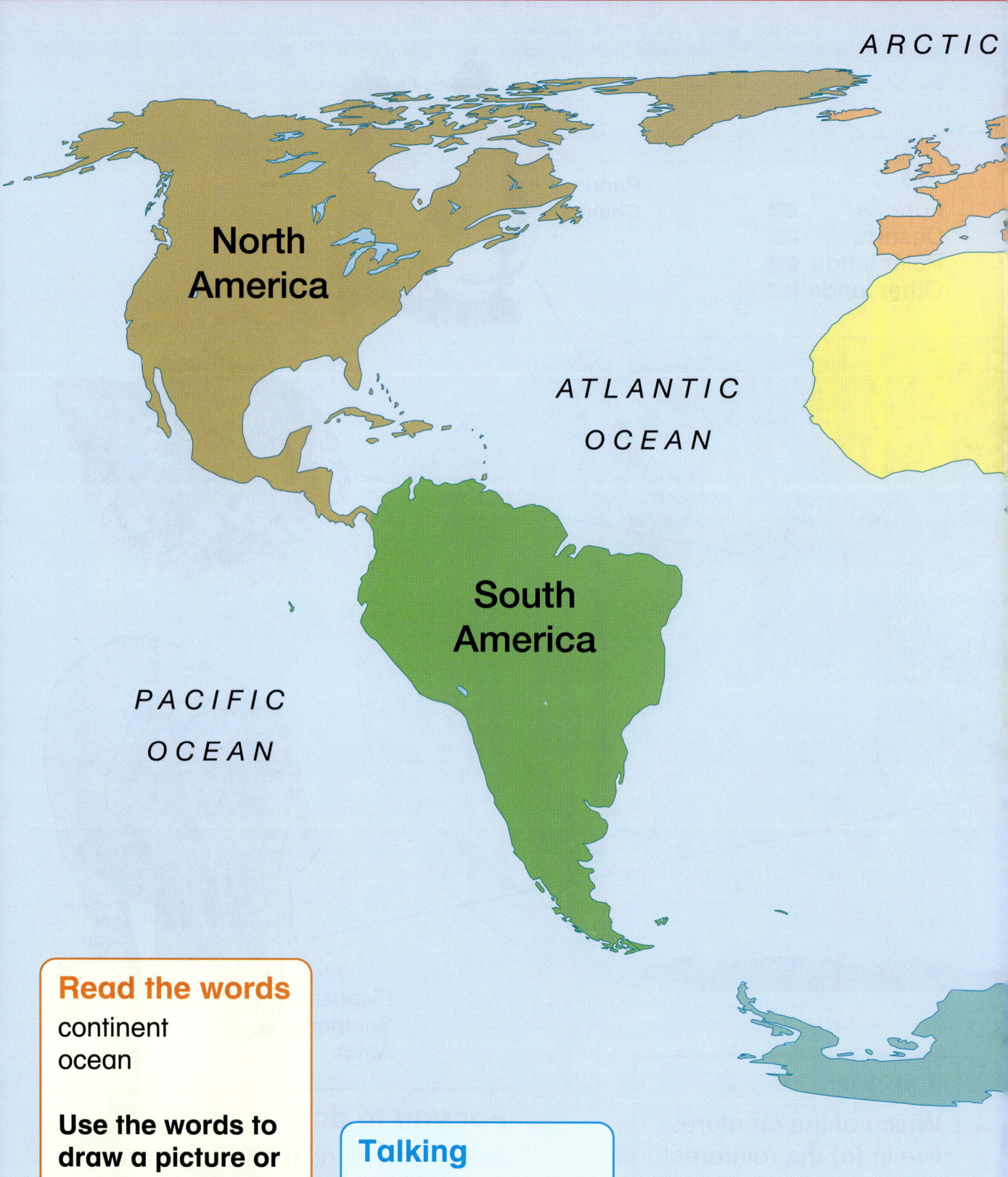

ARCTIC

North America

ATLANTIC OCEAN

South America

PACIFIC OCEAN

Read the words
continent
ocean

Use the words to draw a picture or make a model. Write about it.

Talking
In what way is each continent different?

Amazing fact! There are seven continents and five oceans.

For you to do
Find out if your family or friends have ever lived in another continent. Ask them how it was the same or different to where you live now.

World countries

Read the words

world   countries

Use the words to draw a picture or make a model. Write about it.

Talking
Which country do you think has the most interesting shape?

For you to do
Look at the labels on fruit and vegetables to see where they come from.

Amazing fact! There are about 200 countries in the world.

 United States
 Brazil
 UK
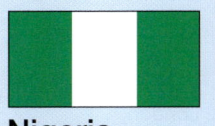 Nigeria
 China
 Australia

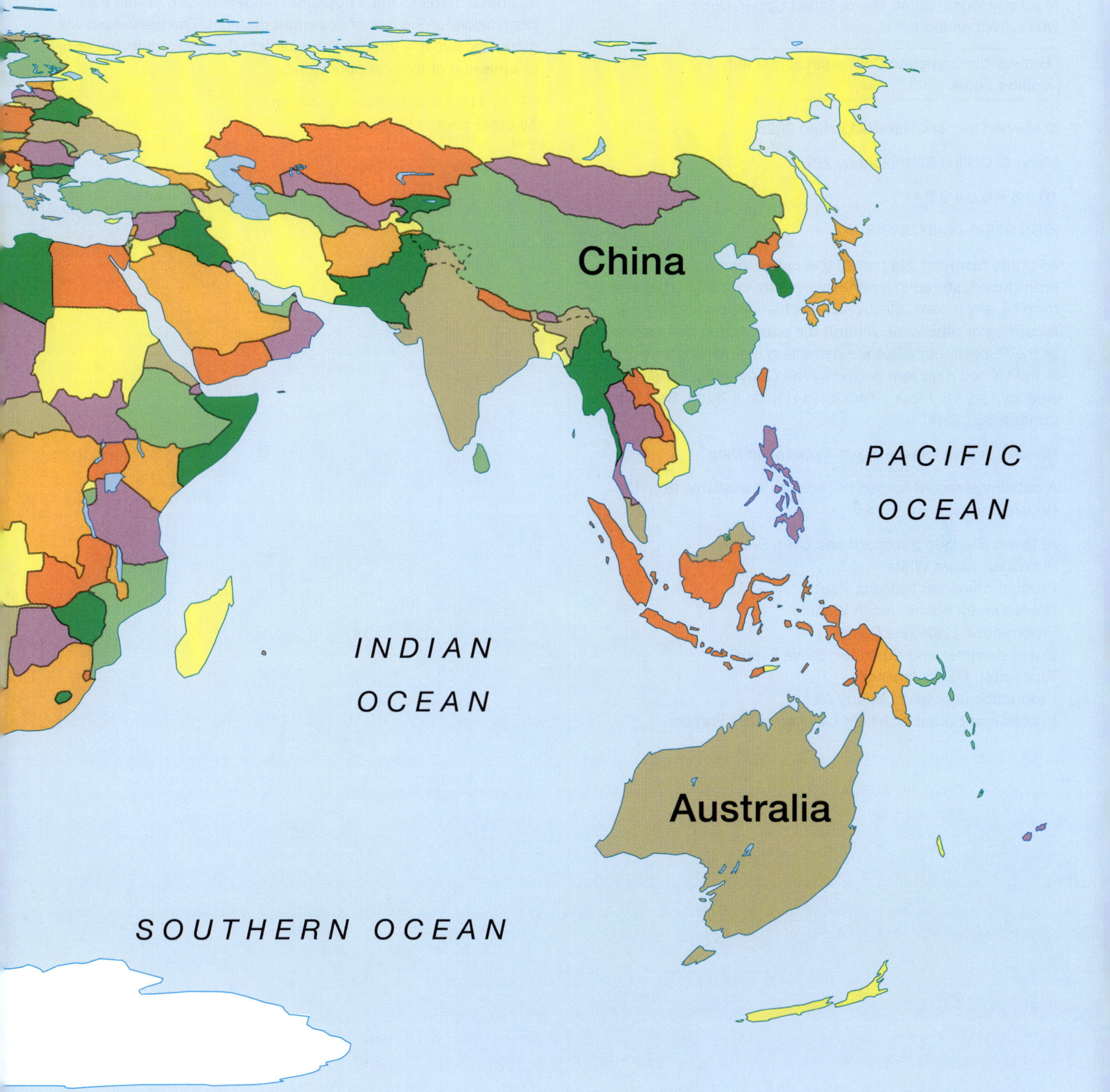

China

PACIFIC OCEAN

INDIAN OCEAN

Australia

SOUTHERN OCEAN

William Collins' dream of knowledge for all began with the publication of his first book in 1819.

A self-educated mill worker, he not only enriched millions of lives, but also founded a flourishing publishing house. Today, staying true to this spirit, Collins books are packed with inspiration, innovation and practical expertise.
They place you at the centre of a world of possibility and give you exactly what you need to explore it.

Published by Collins
An imprint of HarperCollins*Publishers*
The News Building, 1 London Bridge Street, London, SE1 9GF, UK

HarperCollins*Publishers*
Macken House, 39/40 Mayor Street Upper, Dublin 1, D01 C9W8, Ireland

Browse the complete Collins catalogue at
collins.co.uk

© HarperCollins*Publishers* Limited 2025

Maps © Collins Bartholomew 2025

10 9 8 7 6 5 4 3 2 1

ISBN 978-0-00-872829-8

All rights reserved. No part of this publication may be reproduced, stored in a retrieval system, or transmitted in any form by any means, electronic, mechanical, photocopying, recording or otherwise, without the prior written permission of the Publisher or a licence permitting restricted copying in the United Kingdom issued by the Copyright Licensing Agency Ltd, 5th Floor, Shackleton House, 4 Battle Bridge Lane, London SE1 2HX.

British Library Cataloguing-in-Publication Data

A catalogue record for this publication is available from the British Library.

Authors: Stephen Scoffham and Colin Bridge
Publisher: Laura White
Product manager: Natasha Paul
Development editor: Judith Walters
Proofreader: Catherine Dakin
Cover designer and illustrator: Steve Evans
Typesetter: David Jimenez
Production controller: Alhady Ali
Printed and bound in the UK by Martins the Printers

This book is produced from independently certified FSC™ paper to ensure responsible forest management.

For more information visit: www.harpercollins.co.uk/green
collins.co.uk/sustainability

Acknowledgements

The publishers gratefully acknowledge the permission granted to reproduce the copyright material in this book. Every effort has been made to trace copyright holders and to obtain their permission for the use of copyright material. The publishers will gladly receive any information enabling them to rectify any error or omission at the first opportunity.

PP40, 41t, 44t © Stephen Scoffham
All other photos Shutterstock